Vanessa dos Santos de Mello
Isane Vera Karsburg
Bruna Natália V. dos Santos

Heliconias: Tropical beauty

Vanessa dos Santos de Mello
Isane Vera Karsburg
Bruna Natália V. dos Santos

Heliconias: Tropical beauty

Cytogenetics, conservation and genetic improvement

ScienciaScripts

Imprint

Cover image: www.ingimage.com

This book is a translation from the original published under ISBN 978-3-8417-2169-3.

Publisher:
Sciencia Scripts
is a trademark of
Dodo Books Indian Ocean Ltd. and OmniScriptum S.R.L publishing group

120 High Road, East Finchley, London, N2 9ED, United Kingdom
Str. Armeneasca 28/1, office 1, Chisinau MD-2012, Republic of Moldova, Europe
Managing Directors: Ieva Konstantinova, Victoria Ursu
info@omniscriptum.com

Printed at: see last page
ISBN: 978-620-8-60667-1

SUMMARY

CHAPTER 1

INTRODUCTION: GENERAL ASPECTS

Tropical plants and flowers constitute one of the greatest riches of the world's flora. Within this group, heliconia, as the species that make up the genus Heliconia L. are known, are seen as one of the most appreciated groups of herbaceous plants in tropical forests (Berry & Kress 1991; Jorge, 2012). They are commonly known as: bananeira-de-jardim, bananeirinha-de-jardim, bico-de-guará, falsa-ave-do-paraiso, paquevira, and others, depending on the region in which they are found (Sebrae, 2010).

The genus includes a great diversity of species, hybrid varieties and cultivars of ornamental and commercial interest (Berry & Kress, 1991). The plants are perennial, rhizomatous and erect, reaching heights of 3 to 10 metres (Ribeiro et al., 2012). The inflorescences are erect or pendulous (Figure 1), formed by bracts (modified leaves) (Figure 2D) arranged in a dystolic or spiral shape, presenting various colours, sizes, shapes, layouts, textures and numbers, which are the plant structures responsible for arousing its interest and commercial value (Berry & Kress, 1991). The flowers inserted in the bracts are not very conspicuous (Terão et al., 2005) (Figure 2D). They vary in colour from yellow to white, are hermaphrodite, have nectar glands and are pollinated especially by hummingbirds (Berry & Kress, 1991; Marques et al., 2004), Heliconias are propagated by rhizomes (specialised stems) (Figure 2F), which grow horizontally below the surface of the soil and also serve as a source of nutrient and water reserves, making them more resistant to changes in environmental conditions (Rundel et al., 1998). Another way is through seeds, however, germination occurs within 120 days for most species, and some can take up to three years to germinate. This is because the seeds are surrounded by a very hard endocarp (Marques et al., 2004).

Figure 1. Shapes of the inflorescences of different heliconias. A, B and C: Erect; D and E: Pendent.

The name of the genus was established by Linnaeus in 1771, in a reference to the Helicon mountain, located in the region of Beocotia in Greece, where, according to mythological accounts, Apollo and his Muses lived, indicating a kinship between this family and banana trees (genus Musa) (Costa et al., 2011). Heliconias were originally included in the Musaceae family, but later, due to their individual characteristics, they became the sole representative of the Heliconiaceae family (Lamas, 2004).

Figure 2. Vegetative organs of heliconia: (A) plant size; (B) and (C) fruit; (D) yellow flowers inserted in bracts; (E) leaf and (F) rhizome.

Heliconias are plants belonging to the order Zingiberales, with 92 genera and including 8 families: Musaceae; Strelitziaceae; Lowiaceae; Heliconiaceae; Zingiberaceae; Costaceae; Cannaceae and Marantaceae (Figure 3) which comprise monocotyledons, limited to tropical regions, found in woods and understories in humid regions, and are basically differentiated by their flowers and inflorescences (Berry & Kress, 1991; Costa et al., 2011).

Figure 3 - Diagram of the degree of relatedness of heliconia (Heliconiaceae) in the order Zingiberales. Berry & Kress (1991).

According to the phylogenetic analysis carried out on Heliconia accessions using RAPD molecular markers, the Heliconiaceae family was classified as a monophyletic group (Marouelli, 2010). According to the author, the results found corroborate the current classification of families in the Zingiberales order, based on morphological and molecular data. The Musaceae family was found to be the basal family, Strelitziaceae the closest to the heliconia, followed by Marantaceae, Zingiberaceae and Costaceae (Kress et al., 2001; Kress & Specht, 2006). It is suggested that the Heliconiaceae probably underwent rapid irradiation and recent diversification approximately 32 million years ago (Kress & Specht, 2006).

Even though botanical studies have been carried out constantly over the last hundred years, the taxonomy of the Heliconia genus is questionable in terms of determining the exact number of species that occur within the genus (Lamas, 2004). Currently there are an estimated 207 accepted species and 176 registered synonyms (THE PLANT LIST, 2010). In Brazil, around 29 species are accepted and 5 are endemic to the country (Braga, 2015). They are distributed in two regions: the Amazon Basin and the coastal Atlantic Rainforest (Kress, 1990).

The centre of origin of the genus is most likely in north-western South America, a region characterised by high rainfall and nutrient-rich soils (Castro et al., 2007; Kress, 1990), and diversification in the Andes (Colombia and Ecuador) to southern Central America (Panama and Costa Rica), as well as a small group of species distributed in some Pacific Islands (Kress, 1990) (Figure 4).

According to the survey of validated species that make up the genus Heliconia, and the regions of their natural occurrence, Castro et al. (2007) defined that the species and subspecies already described occur in greatest numbers in Colombia (94), followed by Ecuador (60), Panama (56), Costa Rica (47), Brazil (37), Peru (32), Venezuela (26), Nicaragua (22), Guatemala (16), Bolivia (15), Honduras and Mexico (14) and Suriname (13). In Brazil, 34 species are defined as native: H. episcopalis, H. bihai, H. stricta, H. spathocircinata, H. lourteigiae, H. farinosa, H. kautzkiana, H. rivularis, H. sampaioana, H. velloziana, H. chartaceae, H. juruana, H. pendula, H. acuminata, H.angusta, H. psittacorum, H. richardiana, H. aemygdiana, H. pseudoaemygdiana, H. densiflora, H. lasiorachis, H.metallica, H. subulata, H. apparicioi, H. hirsuta, H. marginata, H. latispatha, H. rauliniana, H. julianii, H. rostrata, H. standley, H. tenebrosa, H. timothei and H. velutina (Castro & Graziano, 1997; Castro et al., 2007).

Figure 4 - Worldwide geographical distribution of heliconia species highlighted in red. Mobot (2017).

The species occur at altitudes ranging from 0 to 2,000 metres, although there are some restricted to higher regions. They occur predominantly on the edges of forests and riparian woodlands and in clearings occupied by pioneer vegetation. The vast majority are generally found in humid or flooded regions, but some species can be found in dry areas due to seasonality. They grow in shady places or in full sun, on sandy clay soils (Alonso & Sousa-Silva, 2009).

The genus Heliconia was little known because of the large number of species that exist and the difficulty of collecting and preserving its inflorescences. Since 1985, with the founding of the International Heliconia Society, it has been studied from various angles, making it important for horticulture and landscaping purposes (Ribeiro et al., 2012).

As a result, heliconias have become widely used in landscaping, composing gardens, as well as being used as floral arrangements to decorate environments and events (Costa et al., 2016). Their cultivation is promising in the domestic and foreign markets, as they have characteristics that favour their interest and acceptance, such as: the exotic appearance of the inflorescences and the great variation in colours and shapes, the production of continuous flowers in large quantities, beauty, resistance to transport, post-harvest durability and lower purchase costs (Alonso & Sousa-Silva, 2009; Loges et al., 2005).

In addition to their ornamental importance, they play an important ecological role within ecosystems, acting in the composition of the flora of woodlands and understories, as well as in open environments. In certain ecosystems they act as pioneers in the process of natural regeneration of vegetation and restoration of degraded soil, as well as sustaining coevolutionary relationships with other species of flora and fauna (Lamas, 2004).

Considering the prominence of heliconia species, the great interest they arouse and their growing production and commercialisation, this book will present a broad knowledge of the Heliconia genus, with regard to the number, size and chromosome morphology of the species, studies on pollen viability and meiotic behaviour, collections and germplasm banks for the conservation of the variability of these genetic resources and the prospects for their use in breeding programmes.

Bibliographical references

ALONSO, A. M.; SOUSA-SILVA, J. C. *Heliconia angusta* **Vell.: Characterisation of an**

ornamental plant for cultivation in the Cerrado. Planaltina: Embrapa Cerrados, 2009. 24 p.

BRAGA, J. M. A. 2015. **Heliconiaceae in Lista de Espécies da Flora do Brasil. Botanical Garden of Rio de Janeiro.** Available from: <http://floradobrasil.jbrj.gov.br/jabot/floradobrasil/FB7954>.BF

G. Growing knowledge: an overview of Seed Plant diversity in Brazil. Rodriguésia, v.66, n.4, p.1085-1113. 2015.

BERRY, F.; KRESS, W. J. Heliconia**: an identification guide**. Washington & London: Smithsonia Institution Press, 1991. 334 p.

CASTRO, C. E. F.; GRAZIANO, T.T. Species of the genus Heliconia (Heliconiaceae) in Brazil. **Revista Brasileira de Horticultura Ornamental**, v. 3, p.15-28, 1997.

CASTRO, C. E. F.; GONÇALVES, C.; MAY, A. Updating the nomenclature of species of the genus Heliconia (Heliconiaceae). **Revista Brasileira de Horticultura Ornamental**, v.13, p. 38-62. 2007.

COSTA, F. R. C.; ESPINELLI, F. P.; FIGUEIREDO, F. O. G. **Guia de zingiberales dos Sítos PPbio na Amazônia Ocidental brasileira**. Manaus: Áttema Design Editorial, 2011. 284p.

COSTA, M. G. S.; LEITE, B. S. F.; LOGES, V.; SILVA, E. B. C.; COSTA, A. S.; GUIMARAES, W. N. R.; BRASILEIRO-VIDAL, A. C. Chromosome markers confirm origin of Heliconia hybrids and triploids. **Euphytica (Dordrecht. Online)**, v. 212, p. 1-14, 2016.

JORGE, L. H. A. **Growing Heliconia**. Technical Dossier. SENAI/AM. 2012. 27p.

KRESS, W. J. The diversity and distribution of Heliconia (Heliconiaceae) in Brazil. **Acta Botanica Brasílica**, v 4, n.1, 159-167, 1990.

KRESS, W. J.; PRINCE L. M, HAHN, W. J.; ZIMMER, E. A. Unravelling the evolutionary radiation of the families of the Zingiberales using morphological and molecular evidence. **Systematic Biology**, v. 50, p. 926-944, 2001.

KRESS, W. J.; SPECHT, C. D. The Evolutionary and biogeographic origin and diversification of the tropical monocot order Zingiberales. **Monocots: comparative biology and evolution**, v. 22: 619-630, 2006.

LAMAS, A. M. **Tropical Floriculture: cultivation techniques.** Recife: SEBRAE-PE, 2004. 65p.

LOGES, V.; TEIXEIRA, M. C. F.; CASTRO, A. C. R.; COSTA, A. S. Harvesting, post-harvesting and packaging of tropical flowers in Pernambuco. **Revista Brasileira de Horticultura Ornamental**, v.23, n.3, p.699-702, 2005.

MARQUES, J. M.; COELHO, P. J. A.; FERREIRA, M. A.; AMARAL, Z. P. S.; TORRES, A. C.; AMORIM, J. C.; BUSO, G. S. C. **Study of genetic variability between individuals from populations of** Heliconia bihai **and** Heliconia rostrata. EMBRAPA: Brasília, 2004. 15p.

MAROUELLI, L. P.; INGLIS, P. W., FERREIRA, M. A ; BUSO, G. S. C. Genetic relationships among Heliconia (Heliconiaceae) species based on RAPD markers. **Genetics and Molecular Research**, v.9, p.1377-1387, 2010.

RIBEIRO, W. S.; BARBOSA, W. S.; COSTA, L. C. **Heliconias**. Brasília: Editora Kiron, 2012. 134p.

RUNDEL, P. W.; SHARIFI, M. R.; GIBSON, A. C.; ESLER, K. J. Structural and physiological adaptation to light environment in neotropical Heliconia (Heliconiaceae). **Journal of Tropical Ecology**, v.14, p. 789-801, 1998.

SEBRAE. **Ornamental plants catalogue**. Sebrae: Vale do Ribeira, 2010. 99p.

TERÃO, D.; CARVALHO, A. C. P. P.; BARROSO, T. C. S. F. **Tropical Flowers**. Brasília: Embrapa Informação Tecnológica, 2005

THE PLANT LIST. 2010. Available at: http://www.theplantlist.org/. Accessed on: 08 March 2017.

CHAPTER 2

CHARACTERISATION OF THE CHROMOSOME NUMBER AND STRUCTURE OF THE GENUS HELICONIA L.

Chromosome number

The biological heritage represented by native tropical plants and flowers has little information on their cytogenetic behaviour. For the genus Heliconia, it has been observed that several species have only the chromosome number determined, with insufficient studies on size, morphology, DNA quantification and chromosome banding. The cytogenetic studies available in the literature on the number of chromosomes in Heliconia report two large groups, the first represented by species with 2n=24, characterised as the most frequently observed number within the genus, and the second with 2n=22 chromosomes (Figures 5 and 6). There is also a report of the species H. metallica Planchon & Linden ex Hooker with 2n = 16, 18, 20 and 22, 26 and 36 chromosomes (Agharkar & Bhaduri, 1935). However, this description is discarded when discussing the number of chromosomes that the species of the genus have. Table 1 lists the species that have determined chromosome numbers, as well as their haploid number and ploidy level.

The cytogenetic studies already carried out reveal that the species of the genus are derived from the basic number of x= 11 and 12, showing the predominance of diploid species (2n = 2x = 22 or 2n = 2x = 24), and some triploids (2n = 3x = 36) from natural events. This suggests that aneuploidy (a decrease or increase in the number of chromosome pairs) is responsible for this diversification in chromosome number 2n= 22 and 24.

Heliconias are identified primarily on the basis of morphological characteristics and the colour of their flowers and inflorescences. Significant natural variations are observed within and between species, generated by geographical isolation, environmental and nutritional influences, and have caused confusion among commercial growers and collectors (Berry & Kress, 1991). Furthermore, the genus

Heliconia still presents confusions and uncertainties about the correct classification and the number of existing species (Castro, 1995). In view of this, cytogenetic knowledge about these species combined with morphological, molecular and morpho-agronomic information allows them to be distinguished with greater certainty, contributing to a better taxonomic classification. It also generates data for better planning of genetic improvement, in order to guarantee success in crosses to produce new hybrids.

Table 1. Survey of species in the Heliconiaceae family and their respective chromosome numbers, haploid number and ploidy level.

Species	2n	n	Ploidy level	Reference
H. arabica Lam.	22	11	2x	(Sarkar et al., 1972)
H. *hirsuta* L.f.	22	11	2x	(Dourado et al., 2013)
H. metallica Planchon & Linden ex Hooker	22	11	2x	(Kaemwong & Eksomtramage,1998)
H. wagneriana cv. Rainbow	22	11	2x	(Kaemwong & Eksomtramage, 1998)
H. stricta Huber	22	11	2x	(Miranda et al., 2013)
H. rostrata Ruiz & Pav.	22	11	2x	(Miranda et al., 2013)
H. bihai (L.) L	22	11	2x	(Nascimento et al., 2014)
H. psittacorum L.f.	22	11	2x	(Vieira et al., 2013)
H. brasiliensis Hook.	24	12	2x	(Venkatasubban, 1945)
H. *illustris* W.Bull.	24	12	2x	(Venkatasubban, 1945)
H. latispatha Bentham cv. Distans	24	12	2x	(Andersson, 1984; Costa et al., 2016)
H. imbricata (Kuntze) Baker	24	12	2x	(Andersson, 1984)
H. reticulata (Griggs) Winkl.	24	12	2x	(Andersson, 1984)
H. mariae Hooker fil.	24	12	2x	(Andersson, 1984)
H. bihai (L.) L.	24	12	2x	(Andersson, 1984; Costa et al., 2016)
H. rodriguensis Aristeg	24	12	2x	(Andersson, 1984)
H. latispatha Benth.	24	12	2x	(Andersson, 1984)
H. tortuosa Griggs	24	12	2x	(Andersson, 1984)
H. irrasa R.R. Sm.	24	12	2x	(Andersson, 1984)
H. irrasa cf. ssp. undulata Daniels & Stiles	24	12	2x	(Andersson, 1984)
H. spathocircinata Aristeg.	24	12	2x	(Andersson, 1984; Costa et al., 2016)

H. monteverdensis G.S Daniels & F.G Stiles	24 12	2x	(Andersson, 1984)
H. lourteigiae Emygdio & E. Santos	24 12	2x	(Andersson, 1984)
H. pendula Wawra	24 12	2x	(Andersson, 1984; Costa et al., 2016)
H. marginata (Griggs) Pittier	24 12	2x	(Andersson, 1984)
H. pogonantha var. holeryrhra G.S Daniels & F.G Stiles	24 12	2x	(Andersson, 1984)
H. pogonantha var. veraguasensis W.J. Kress	24 12	2x	(Andersson, 1984)
H. ramonensis Daniels & A.H Gentry	24 12	2x	(Andersson, 1984)
H. xanthovillosa W.J. Kress	24 12	2x	(Andersson, 1984)
H. nigripraefixa Dodson & Gentry	24 12	2x	(Andersson, 1984)
H. chartacea Lane ex Barreiros	24 12	2x	(Andersson, 1984)
H. riopalenquensis Dodson & A.H. Gentry	24 12	2x	(Andersson, 1984)
H. nutans Woodson	24 12	2x	(Andersson, 1984)
H. acuminata L. C. Richard.	24 12	2x	(Andersson, 1984)
H. psittacorum L.f.	24 12	2x	(Andersson, 1984; Costa et al., 2016; Kaemwong & Eksomtramage, 1998)
H. richardiana Miquel	24 12	2x	(Andersson, 1984)
H. densiflora B. Verlot	24 12	2x	(Andersson, 1984)
H. longiflora R.R. Smith	24 12	2x	(Andersson, 1984)
H. venusta Abalo & G. Morales	24 12	2x	(Andersson, 1984)
H. aemygdiana Burle-Marx	24 12	2x	(Andersson, 1984)
H. virginalis Abalo & Morales	24 12	2x	(Andersson, 1984)
H. rostrata Ruiz & Pavón	24 12	2x	(Costa et al., 2016; Kaemwong & Eksomtramage, 1998)
H. caribaea Lamarck cv. Purpurea	24 12	2x	(Kaemwong & Eksomtramage, 1998)
H. latispatha Bentham cv. Red-Yellow Gyro	24 12	2x	(Kaemwong & Eksomtramage, 1998)
H. psittacorum L.f. x *H. marginata* (Griggs) Pittier	24 12	2x	(Kaemwong & Eksomtramage, 1998)
H. psittacorum L.f. cv. Rubra	24 12	2x	(Kaemwong & Eksomtramage, 1998)
H. *psittacorum* L.f. cv. Parakeet	24 12	2x	(Kaemwong & Eksomtramage, 1998)
H. *psittacorum* L.f. cv. Lady Di	24 12	2x	(Kaemwong &

			Eksomtramage, 1998; Lee et al., 1984).
H. psittacorum L.f. cv. Sassy	24 12	2x	(Kaemwong & Eksomtramage, 1998)
H. stricta Huber cv. Jamaican dwarf	24 12	2x	(Kaemwong & Eksomtramage, 1998)
H. psittacorum L.f. x *H. spathocircinata* Aristeg. cv. Golden Torch	24 12	2x	(Costa et al., 2016; Kaemwong & Eksomtramage, 1998)
H. psittacorum L.f. x *H. spathocircinata* cv. Golden Torch Adrian	24 12	2x	(Costa et al., 2016; Kaemwong & Eksomtramage, 1998)
H. psittacorum L.f. cv. Tay	24 12	2x	(Lee et al., 1994)
H. psittacorum L.f. cv. Andromeda	24 12	2x	(Lee et al., 1994)
H. psittacorum L.f. cv. Lady Di	24 12	2x	(Lee et al., 1994)
H. stricta Huber cv. Fire Bird	24 12	2x	(Guangsui et al.,2010)
H. orthotricha cv. Lehua	24 12	2x	(Guangsui et al., 2010)
H. psittacorum cv. nickeriensis	24 12	2x	(Guangsui et al., 2010)
H. caribaea Lamarck x *H.bihai* cv. Hot Rio Nites	24 12	2x	(Guangsui et al., 2010)
H. stricta cv. Iris Bannochie	24 12	2x	(Guangsui et al., 2010)
H. collinsiana x *H. bourgeana* cv. Pedro Ortiz	24 12	2x	(Guangsui et al., 2010)
H. caribaea Lamarck x *H.bihai* (L.) L. cv Jacquinii	24 12	2x	(Guangsui et al., 2010)
H. stricta Huber	24 12	2x	(Costa et al., *2016)*
H. wagneriana Petersen	24 12	2x	(Costa et al., 2016)
H. psittacorum L.f. cv. Red Opal	24 12	2x	(Costa et al., 2016)
H. psittacorum L.f. cv. Strawberries & Cream	24 12	2x	(Costa et al., 2016)
H. collinsiana Griggs	24 12	2x	(Costa et al., 2016)
H. rauliniana Barreiros	24 12	2x	(Costa et al., 2016)
H. caribaea Lamarck X *H. bihai* (L.) L. cv. Jacquinii	24 12	2x	(Guangsui et al., 2010; Costa et al., 2016)

H. caribea	24 12	2x	(Mahanty, 1970)
H. aurantiaca	24 12	2x	(Mahanty, 1970)
H. rubra	26 13	2x	(Venkatasubban, 1945)
H. *densiflora* Verlot cv. Fire Flash	36 12	3x	(Kaemwong & Eksomtramage, 1998)
H. psittacorum L.f. cv. Petra	36 12	3x	(Lee et al., 1994)
H. psittacorum L.f. cv. Iris	36 12	3x	(Lee et al., 1994)
H. *psittacorum* L.f. cv. Sassy	36 12	3x	(Lee et al., 1994)
H. *psittacorum* L.f. cv. Sassy	36 12	3x	(Lee et al., 1994; Costa et al., 2016)
H. *psittacorum* L.f. cv. Suriname Sassy	36 12	3x	(Costa et al., 2016)

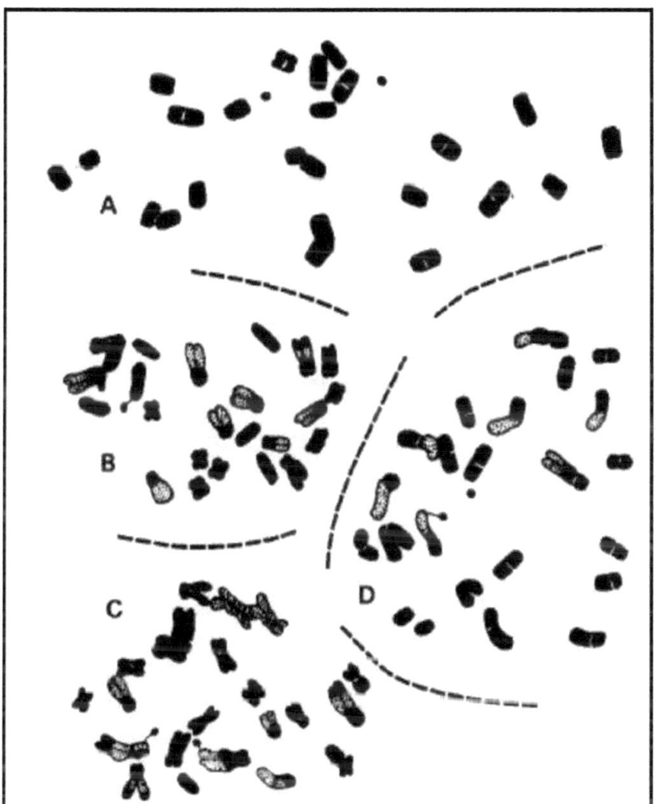

Figure 5: Heliconia chromosomes: (A) *H. Pogonantha* var. *Veraguasensis;* (B) *H. Riopalenquensis;* (C) *H. sparhooircinata;* (D) *H. psittacorum.* Andersson (1984).

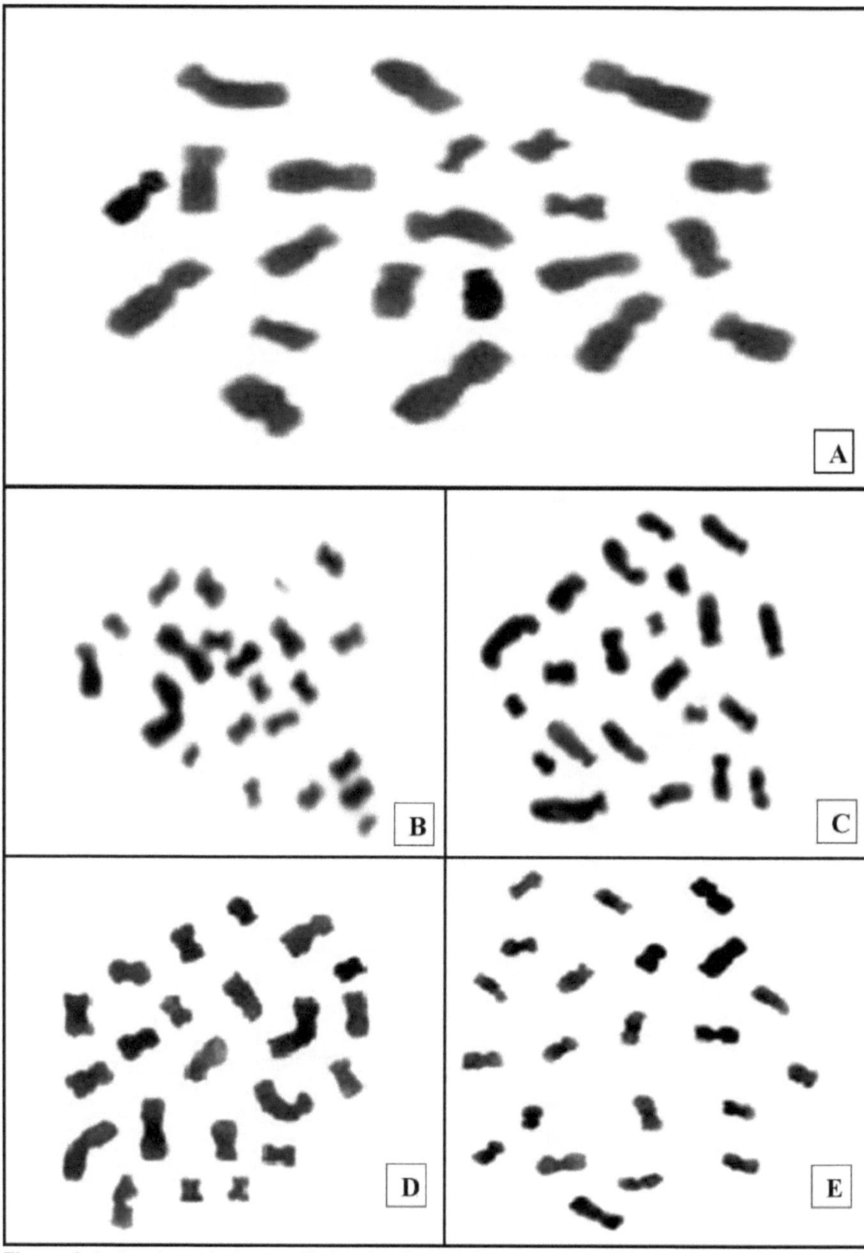

Figure 6. heliconia metaphase, all showing **2n=22** chromosomes: (A) *H. stricta;* (B) **H.** *bihai;* (C) **H.** *hirsuta;* (D) **H.** *psittacorum,* and (E) **H.** *rostrata.* Dourado et al. (2013); Miranda et al. (2013); Nascimento et al. (2014); and Vieira et al. (2013).

Size and morphology in terms of centromere position

With regard to the chromosome morphometry of the species already studied, their karyotypes are made up of metacentric, submetacentric and some species have acrocentric chromosomes (Figure 7). In terms of size, they are considered to be short to medium.

Table 2 shows the species whose morphology has been described in terms of centromere position and chromosome size. In addition to these, Andersson (1984) in 31 heliconia taxa, observed chromosome size ranging from 0.7 to 1.7μm, classified as short and medium, with an increase in karyotypic symmetry as chromosome length increased. The satellites were found attached to the longest arms of one of the larger pairs (Andersson, 1984). According to the same author, the good taxonomic distribution of heliconia plant material allows us to conclude that morphological chromosomal variation is not of great importance for the diversification of the genus, and this applies to both structural and numerical characteristics.

Table 2. Karyotypic formula and chromosome length (μm) of heliconia species.

Species	Karyotypic formula	Length (Mm)	Reference
H. *bihai* (L.) L	7M + 4SM 12SM 6A+3M+3SM	0.67 - 3.96 2.71 - 1.02 1.40 - 4.50	(Nascimento et al., 2014) (Costa et al., 2008; Costa et al., 2016) (Mahanty, 1970)
H. *stricta* Huber	9SM+2M	-	(Miranda et al., 2013)
H. *rostrata* Ruiz & Pavón	5SM+5M+1A	-	(Miranda et al., 2013)
H. *hirsuta* L.f.	6SM+4M+1A	0.28 - 4.36	(Dourado et al., 2013)
H. *psittacorum* L.f.	4SM+6M+1A	1.19 - 3.29	(Vieira et al., 2013)
H. orthotricha *cv.* Lehua	12M	-	(Guangsui et al., 2010)
H. *psittacorum* cv. Nickeriensis	5M+7SM	-	(Guangsui et al., 2010)
H. *caribaea* Lamarck x *H.bihai* cv. Hot Rio Nites	5M+7SM	-	(Guangsui et al., 2010)
H. *stricta cv.* Iris	8M+4SM	-	(Guangsui et al., 2010)

Bannochie			
H. *collinsiana* x bourgeana cv. Pedro Ortiz	5M+7SM	-	(Guangsui et al., 2010)
H. *caribaea* Lamarck x *H.bihai* (L.) L. cv Jacquinii	2M+12SM	-	(Guangsui et al., 2010)
H. *psittacorum* cv. Red Opal	-	1.15 - 2.64	(Costa et al., 2016)
H. psittacorum cv. Strawberries & Cream	-	1.25 - 3.35	(Costa et al., 2016)
H. psittacorum L.f.	-	1.14 - 2.73	(Costa et al., 2016)
H. spathocircinata Aristeg	-	1.25 - 3.02	(Costa et al., 2016)
H. spathocircinata x *H. psittacorum* cv. Golden Torch	-	1.44 - 2.98	(Costa et al., 2016)
H. spathocircinata x *H. psittacorum* cv. Golden Torch Adrian	-	1.13 - 2.89	(Costa et al., 2016)
H. caribea x H. bihai cv. Jacquinii	-	1.18 - 2.57	(Costa et al., 2016)
H. psittacorum cv. Suriname Sassy	-	0.88 - 2.39	(Costa et al., 2016)
H. psittacorum cv. Sassy	-	1.12 - 2.53	(Costa et al., 2016)

(-) Species that do not have a determined karyotype formula or chromosome length.

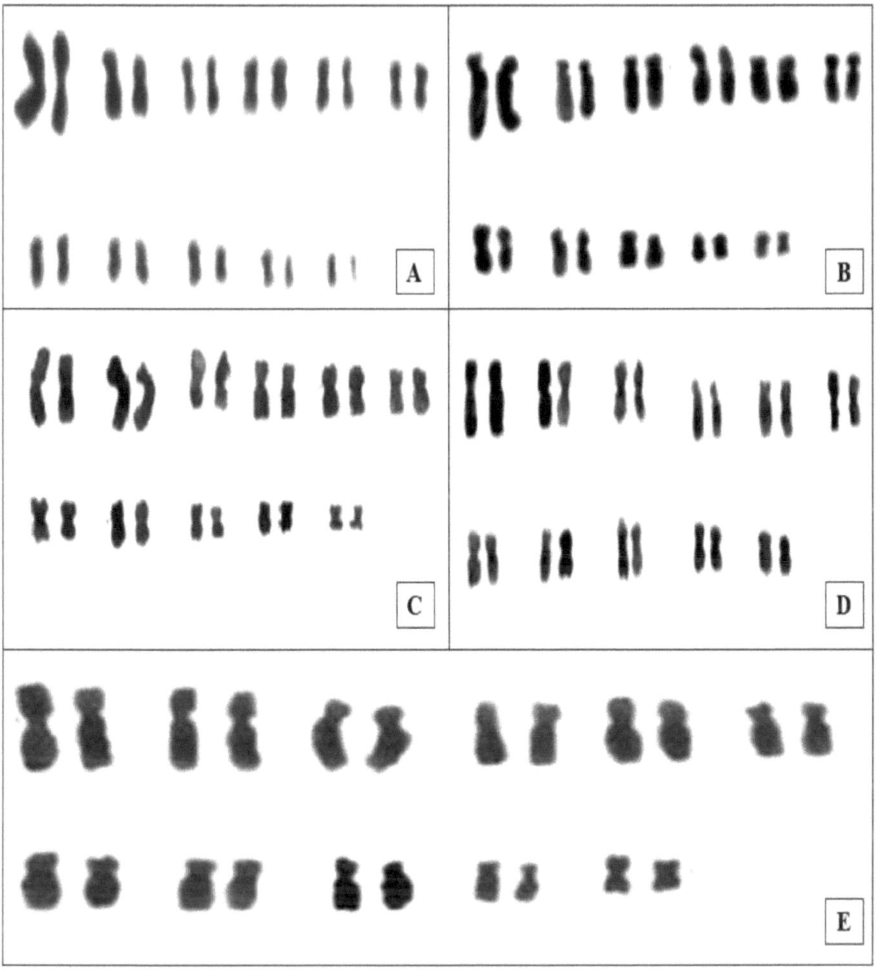

Figure 7. *Heliconia* chromosome sizes: (A) *H. bihai;* (B) *H. hirsuta*; (C) *H. psittacorum;* (D) *Heliconia rostrata;* and (E) **H.** *stricta.* Dourado et al. (2013); Miranda et al. (2013); Nascimento et al. (2014); and Vieira et al. (2013).

Chromosome banding

Chromosome banding techniques have been commonly used in cytogenetics (Guerra, 2000). Among these is silver nitrate staining, which reveals the position of the nucleolus organising region (NOR) (Howell & Black, 1980; Sumner, 2003). Differences in the Ag-NOR band pattern, combined with other cytogenetic techniques, can be used to characterise genetic resources and better understand the evolution of species' genomes, especially when it comes to studying polyploidy in the evolution of plant genomes (Schifino-Wittmann, 2004; Sumner, 2003; Wendel, 2000). However, as is well known, most cytogenetic studies related to heliconia are restricted to the use of conventional techniques for determining chromosome number and morphometry. Ag-NOR banding was only carried out for the species H. bihai (L.) L., allowing the nucleolus organising regions (NOR) to be identified and located in the middle portion of the centromere, in chromosome pairs 1 and 2, which confirmed the diploid behaviour of this species (Nascimento et al., 2014) (Figure 8 and 9).

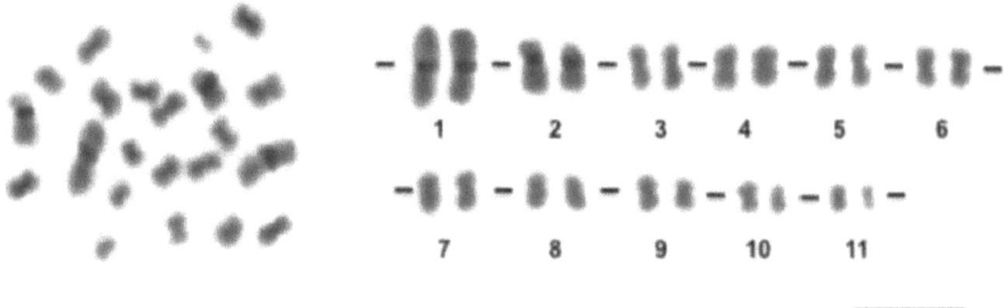

Figure 8. Chromosomes stained with AgNO3 for 18 hours at 35°C with the presence of active NOR in chromosome pairs 1 and 2. Bar=5pm. Nascimento et al. (2014).

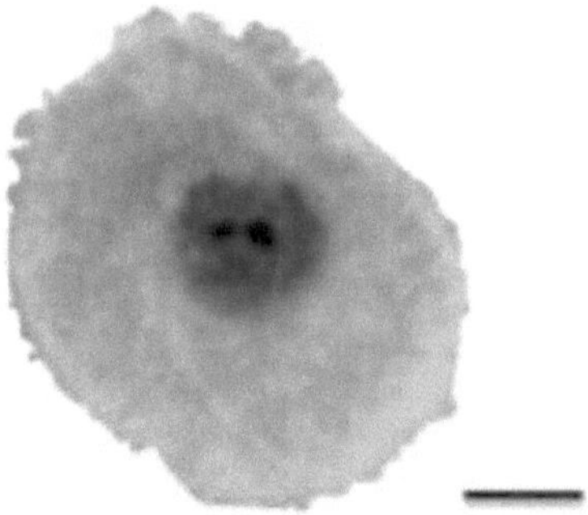

Figura 9. Cell in interphase stained with AgNO3 for 18 hours at 35°C. Presence of two nucleoli. Bar = 5 pm. Nascimento et al. (2014).

With more refined banding, Costa et al. (2008) compared the distribution pattern of CMA3 and DAPI bands and the location of DNAr 5s and 45s sites in nine Heliconia accessions, which had diploid numbers of 2n=24, a satellite pair, a gradual karyotype and semi-reticulate interphase nuclei, and observed a karyotypic differentiation of the accessions analysed by the location of the different chromosome marks (Figure 9 and

10).

The lack of this information makes it difficult to understand the karyotypic evolution of the genus and its phylogenetic relationships, and although many species of heliconia have the same chromosome number, banding can show karyotypic differences that are not detected by classical cytogenetics.

Costa et al. (2016) studied the cytogenetic behaviour of 18 Heliconia genotypes using the CMA and DAPI fluorochrome staining techniques and FISH for 45S and 5S DNAr sites and observed 16 diploid genotypes 2n = 2x = 24 and 2 triploids 2n = 3x = 36 chromosomes. Using these techniques, a CMA+/DAPI0/DNAr 5S band was found in the proximal region of two and three small chromosomes in the diploid and triploid genotypes, respectively. The CMA++/DAPI-/DNAr 45S markings varied from two to four sites per genotype, with some heteromorphisms observed. Based on the data found, it was possible to confirm the origin of the triploid genotypes and natural hybrids. Figures 10, 11 and 12 show the data found.

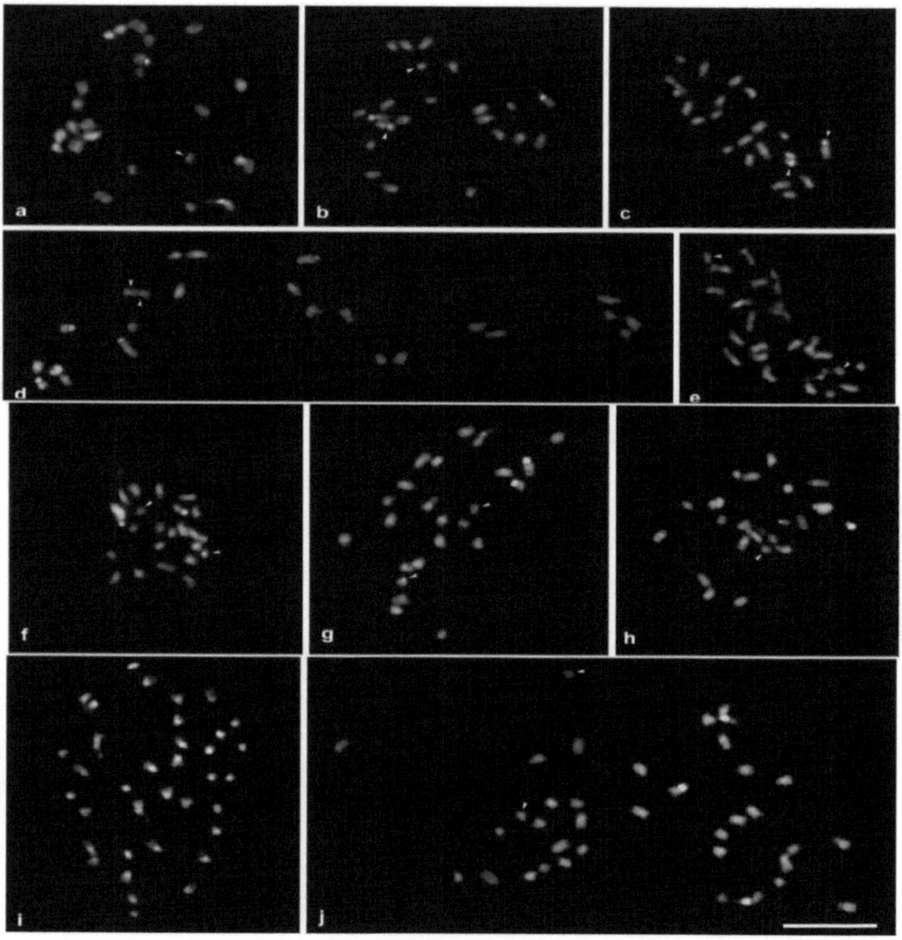

Figura 10. Metaphases of *Heliconia* genotypes stained with CMA (yellow) and DAPI (grey): (A) H. *psittacorum* cv. 'Strawberries & Cream'; (B) H. *Psittacorum;* (C) H. *psittacorum* cv. 'Red Opal'; (D) H. *spathocircinata;* (E) H. *bihai;* (F) H. *psittacorum* x H. *Spathocircinata* cv. 'Golden Torch'; (G) H. *psittacorum* x H. *Spathocircinata* cv. 'Golden Torch Adrian'; (H) H. *caribaea* x *H. Bihai* cv. 'Jacquinii'; (L) *H. psittacorum* cv. 'Suriname Sassy'; (J) *H. Psittacorum* cv. 'Sassy'. The arrowheads point to the CMA+/ DAPI- Bands co-localised with 5S rDNA. Bar corresponds to 10 um. Costa et al. (2016).

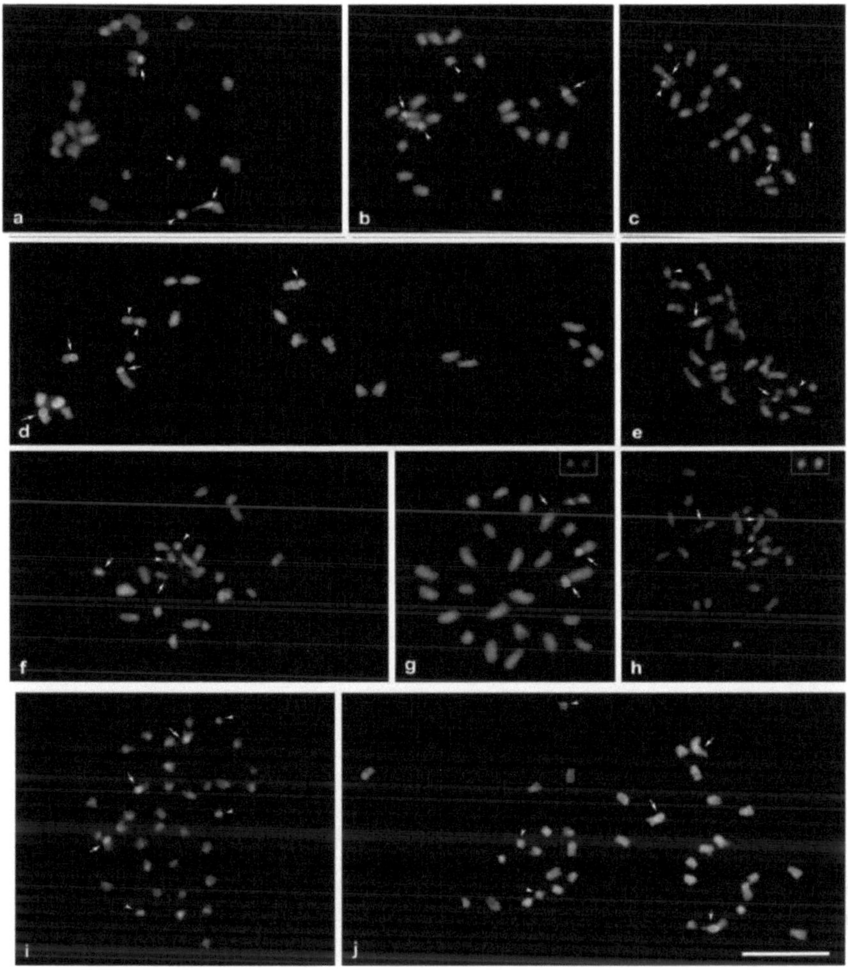

Figura 11. Metaphases of *Heliconia* genotypes hybridised in situ with 45S rDNA (green) and 5S rDNA probes (red): (A) H. *psittacorum* cv. 'Strawberries & Cream'; (B) H. *psittacorum;* (C) H. *psittacorum* cv. 'Red Opal'; (D) H. *spathocircinata;* (E) H. *bihai;* (F) *H. caribaea* x *H. bihai* cv. Jacquinii; (G) *H. psittacorum* x *H. spathocircinata* cv. 'Golden Torch'; *H Psittacorum* x *H. spathocircinata cv.* 'Golden Torch Adrian'; (L) *H. psittacorum cv.* 'Suriname Sassy', (J) *H. psittacorum cv.* 'Sassy' Chromosomes in continuous boxes (G, H) correspond to chromosomes with 5S rDNA sites belonging to other cells. Arrowheads point to 5S RNA sites and arrows to 45S rDNA sites. Bar corresponds to 10 pm. Costa et al. (2016).

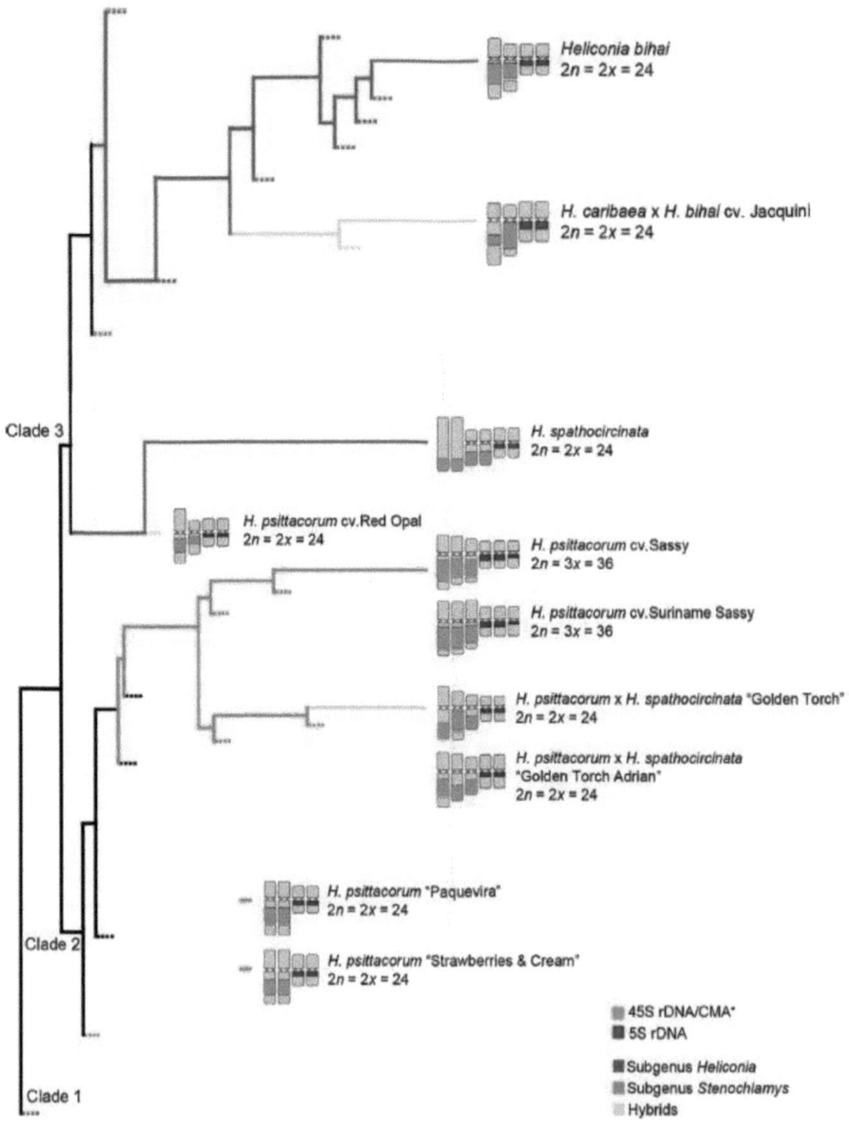

Figura 12. Chromosomal positions of 45S and 5S rDNA sites for *Heliconia* genotypes, mapped on the phylogenetic tree proposed for the genus (modified from Marouelli et al., 2010). Blocks in green indicate 45S rDNA sites co-localised with CMA+/ DAPI- bands; blocks in red indicate 5S rDNA sites. Costa et al. 2016.

Bibliographical references

ANDERSSON, L. The chromosome number of Heliconia (Musaceae). **Nordic Journal**

of Botany, v.4, n.2, p.191-194, 1984.

AGHARKAR, S. P.; BHADURI. P. N. Variation in chromosome numbers in Musaceae. **Current Sciense**, v.3, p.615-617, 1935.

BERRY, F.; KRESS, W. J. Heliconia**: an identification guide**. Washington & London: Smithsonia Institution Press, 1991. 334 p.

CASTRO, C. E. F. Interrelationships of the Zingiberales families. **Revista Brasileira de Horticultura Ornamental**, v.1, n.1, p.2-21, 1995.

COSTA, A; FERREIRA-LEITE, B. S.; LOGES, V.; BERNARDES, E. C. S.; BRASILEIRO-VIDAL, A. C. Distribution pattern of CMA3 bands and localisation of 5S and 45S DNAr sites in the analysis of Heliconia (Heliconiaceae) accessions. In: CONGRESSO BRASILEIRO DE GENÉTICA, 54, 2008, Salvador. **Proceedings...** Salvador: Brazilian Genetics Association, 2008. 54, p.203.

COSTA, M. G. S.; LEITE, B. S. F.; LOGES, V.; SILVA, E. B. C.; COSTA, A. S.; GUIMARAES, W. N. R.; BRASILEIRO-VIDAL, A. C. Chromosome markers confirm origin of Heliconia hybrids and triploids. **Euphytica (Dordrecht. Online)**, v. 212, p. 1-14, 2016.

DOURADO, J.; OLIVEIRA, A. N. B.; VIEIRA, A.; KARSBURG, I. V. Chromosomal characterisation of Heliconia hirsuta L. F. **Magistra**, v. 25, p.193-195, 2013.

GUANGSUI, Y.; JUNMEI, Y.; QINGYUN, L.; YANXIA, Z.; ZHIQUN, Z. Karyotypes comparison of eight varieties of Heliconia spp. **Chin J Trop Crops**, v. 12, p. 2118-2123, 2010.

GUERRA, M. Patterns of heterochromatin in plant chromosomes. **Genetics and**

Molecular Biology, v. 23, n. 4, p.1029-1041, 2000.

HOWELL, W. M.; BLACK, B. A. Controlled silver-straining of nucleolus organiser regions with a protective colloidal developed:a 1-step method. **Experientia**, v. 36, p.1014-1015, 1980.

KAEMWONG, S.; EKSOMTRAMAGE, L. Chromosome numbers of genus Heliconia. Songklanakarin **Journal of Science and Technology**, v. 20, p.489-495, 1998.

LEE, Y. H.; NG, N. Y.; GOH, C. J. Pollen formation and fruit set in some cultivars of Heliconia psittacorum. **Scientia Horticulturae**, v. 60, p.167-172, 1994.

MAHANTY, H. K. A cytological study of the Zingiberales with special reference to their taxonomy. **Cytologia**, v.35, p.1349, 1970.

MIRANDA, D. P.; OLIVEIRA, A. N. B.; MELLO, V. S. VIEIRA, A.; KARSBURG, I. V. Chromosomal characterisation of two species of the genus Heliconia. **Magistra**, v. 25, p.192-192, 2013.

NASCIMENTO, H. R.; GALLO, R.; KARSBURG, I. V.; OLIVEIRA, A. S. Cytogenetic and identification of the nucleolus organiser region in Heliconia bihai (L.) L. **Revista Ceres**, v. 61, n.4, p. 451-457, 2014.

SCHIFINO-WITTMANN, M. T. Polyploidy and its impact on the origin and evolution of wild and cultivated plants. **Revista Brasileira Agrociência**, v.10, n. 2, p. 151-157, 2004.

SARKAR, A. K.; DATTA, N.; CHATTERJEE, U. Chromosome survey of certain angiosperms. **Bulletin of the Botanical Survey of Indica**, v. 14 p.170, 1972.

SUMNER, A. T. **Chromosomes: Organisation and function**. Malden, Blackwell. 2003. 286p.

VENKATASUBBAN, K. R. A preliminary survey of chromosome numbers in Scitaminae of Bentham and Hooker. - Proc. Indian Acad. **Science Series B**, v.23, p. 281-300, 1945.

VIEIRA, A.; OLIVEIRA, A. N. B.; MIRANDA, D. P.; KARSBURG, I. V. Chromosome morphometry of Heliconia psittacorum. **Magistra**, v. 25, p. 366-367, 2013.

WENDEL, J. F. Genome evolution in polyploids. **Plant Molecular Biology**, v. 42, p. 225-249, 2000.

CHAPTER 3

POLLEN VIABILITY OF HELICONIA

A plant breeding programme is based on obtaining superior cultivars through genetic manipulation of the germplasm of the species of interest. The success of these programmes requires a number of important factors, including the most efficient selection of the genetic resources present in these materials and crossbreeding, where both between varieties and cultivars of a species and between species directly depends on the viability of the pollen grain (Techio et al., 2006).

Thus, taking into account the role of the pollen grain in the life cycle of a plant, one way of estimating pollen viability would be to use the pollen for pollination and then analyse fruit set. However, this methodology is time-consuming or impossible to carry out, so other methods are often used to assess pollen grain viability (Karakaya, 2011), such as histochemical tests (lugol 2%, acetic orcein 2%, Alexander's reagent, Astra blue 1%, etc.) and in vitro germination, in which a culture medium is used to germinate the pollen grains to determine their viability, both of which provide quick and inexpensive results.

Some species of heliconia have already been studied through histochemical tests using various dyes and pollen germination in vitro. In Heliconia rostrata Ruiz & Pav. the species showed average percentages of pollen viability for: lugol 2% (59.26%), fuchsin 0.5% (59.32%), malachite green 1% (64.73%), astra blue 1% (64.79%) and acetic orcein 2% (65.53%), with the sole exception of alexander's reagent which showed a high percentage of pollen grain viability, with 78.79% for the species. The average percentage of pollen tube germination was high at 74.93%, in line with the pollen viability value found with Alexander's reagent (Mello et al., 2016a) (Figure 13). For this same species, using the dye 2,3,5 chloride-triphenyltetrazolium at concentrations of 0.075% and 0.30% at 6:00, 12:00, 18:00 and 24:00 hours of exposure showed high pollen viability at 12:00 and 18:00 hours, with 96.75% and 95.80%, respectively (Mello et al., 2016b). The salts of 2,3,5-triphenyl tetrazolium chloride (TTC) confirm pollen viability through the reaction of active enzymes that cause the colour of the grains to change to red in the presence of oxidative activity. The reaction of the dye indicates the presence of enzymatic activity in living cells (Vieitez, 1952; Stanley, 1965). The application of this differentiated dye to determine the pollen viability of species is necessary, particularly in cases where dyes are applied that may overestimate the percentage of pollen grain viability. Thus, the data generated by the TTC test generates more reliable results, since it colours according to the cellular respiration of viable pollens.The pollen viability estimated through the germination of the pollen tube of Heliconia rostrata was determined using the liquid medium prepared according to Trabelsi (1985) and a 10% sucrose solution with 0.001% boric acid (H_3BO_3). The material remained at a temperature of 32°C in a germination chamber with a photoperiod. After 48 hours, the material was analysed and an average germination percentage of 74.93% was observed (Figure 12 N and O), which is considered high, as a viable pollen should have 50 to 80% germinated grains with well-developed tubes. For a high germination rate, it is necessary to understand a series of factors such as temperature, pH, adjustments to the composition of the culture medium and the pollen tube emission period (Zambom, 2014), factors which differ according to the requirements of each species (Mello et al., 2016a).

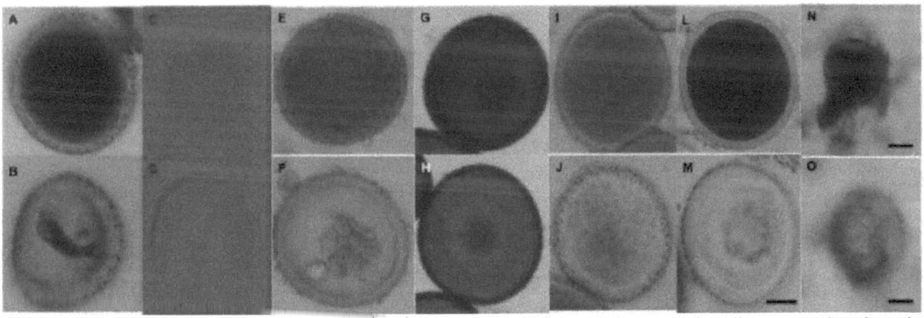

Figure 13. Pollen grains of *H. rostrata* stained with different dyes and pollen germination. Lugol 2% A) viable B) non-viable; Fuchsin 0.5% C) viable D) non-viable; Malachite green 1% E) viable F) non-viable; Astra blue 1% G) viable H) non-viable; Acetic orcein 2% I) viable J) non-viable; Alexander's Reactive L) viable M) non-viable. **Bar = 10pm. N) germinated pollen grain; O) non-germinated pollen grain. Bar = 20 pm. Mello et al. (2016).**

In several species of heliconia, pollen viability was assessed using Alexander's reagent. The species *H. subulata* (Figure 14C) showed the highest viability (97.63%), followed by: *H. velerigera* (Figure 14E) (89.48%), *H. episcopalis* (85.40%), *H flabellata* (83.55%), *H. latispatha* (83.03%), *H. lingulata* (77.68%) (Figure 14D), *H. psittacorum II* (76.43%), therefore considered the most fertile. *H. psittacorum I* showed average viability (35.51%), while *H. collinsiana* (13.57%), *H. rivulares* and *H. raulineana* (6.73%) (Figure 14 F) were low, and *H. hybrid* torch (Figure 14B), *H. imbricata and H. orange* torch 100% sterile (Franco et al., 2012).

In three accessions of *Heliconia psittacorum* collected in different locations, Nascimento et al. (2016) found a high percentage of pollen viability, ranging from 84.00% to 99.50%, which made it possible to infer that the gamete has a high capacity for fertilisation.

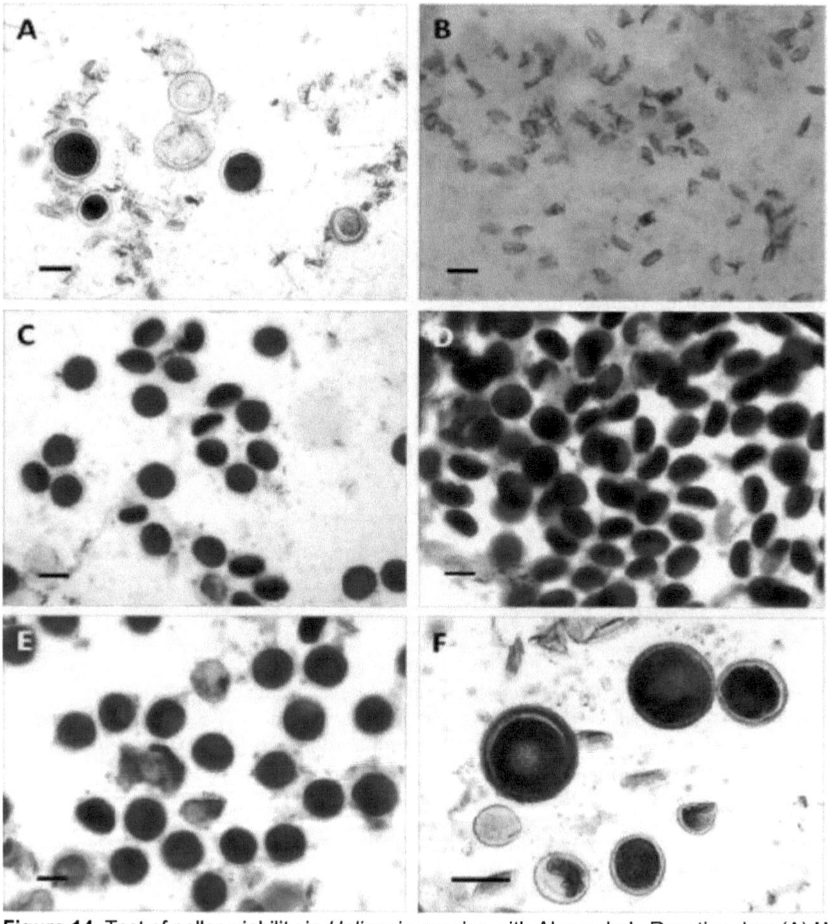

Figure 14. Test of pollen viability in *Heliconia* species with Alexander's Reactive dye: (A) **H.** *collinsiana;* (B) **H.** *hybrid tortoise;* (C) **H.** *subulata;* (D) **H.** *lingulata;*

(E) *H. velerigera;* (F) *H. rauliniana.* **Bar= 50 pm. Franco et al. (2012).**

These pollen viability studies provide information on the reproductive capacity of these species and show that the vast majority of the heliconia species studied are capable of reproducing sexually. This type of information is essential for a genetic improvement programme, highlighting once again the importance of characterising the genetic resources of these materials.

Bibliographical references

FRANCO, V. R.; ZUCCARI, L. D. A.; PINTO-MAGLIO, C. A. F.; CASTRO, C. E. F. Meiotic behaviour and pollen viability in Heliconia pogonantha and Heliconia psittacorum. In: CONGRESS INTERINSTITUTIONAL SCIENTIFIC INITIATION, 6, 2012, Jaguariúna. **Proceedings...** Jaguariúna: CIIC, 2012. p1-10.

KARAKAYA, D. Effects of inflorescence on pollen viability and morphologyof Strawberry (Fragaria vesca L.). **Journal of Science and Technology**, v. 1, p. 43-47, 2011.

MELLO, V. S.; SANTOS, B. N. V.; DAMASIO, J. F.; MACEDO, W. A.; LEITE, D. M.; KARSBURG, I. V. Viability and pollen germination in marsh banana. **Ciência & Tecnologia: Fatec-JB**, v. 8, n.1, 2016a.

MELLO, V. S.; SANTOS, B. N. V.; MACEDO, W. A.; DAMASIO, J. F.; KARSBURG, I. V. Estimation of the pollen viability of Heliconia rostrata L. (Heliconiaceae) by means of the Tetrazolium colourimetric test. In: SEMINAR ON BIODIVERSITY AND AMAZON AGROECOSYSTEMS. **Proceedings...** Unemat: Alta Floresta, p.693-697, 2016b.

NASCIMENTO, T. O.; SILVA, P. C.; SILVA, C. G.; BOTINI, A. F.; FRANÇA, R. P. A.; SANTOS, J. L.; SILVA, C. A. Floral durability, pollen viability and stigmatic receptivity in Heliconia psittacorum (HELICONEACEAE). **Scientific Electronic Archives**, p.316-319, 2016.

STANLEY, R. G. Physiology and uses of tree pollen. **Agricultural Science Review**, v.3, p.9-17, 1965.

TECHIO, V. H.; DAVIDE, L. C.; PEDROZO, C. A.; PEREIRA, A. V. Pollen grain viability of accessions of elephant grass, millet and interspecific hybrids (elephant grass x millet). **Acta Scientiarum Biological Sciences,** v. 28, n. 1, p. 7-12, 2006.

TRABELSI, M. A. A rebliable method for testing fruit setting ability in tomato using "in *vitro* pollen germination. **Mededelingen van de faculteit Landbouwwetenschappen Rijksuniversiteit Gent (Belgium)**, v.50, n.4, p.1343-1356, 1985.

VIEITEZ, E. El uso del Chloruro 2,3,5- trifeniltetrazolium para determinar la vitalidad del polen. **Anales de Edafología y Fisiologia Vegetal,** v.11, p.97- 308, 1952.

ZAMBON, C. R.; SILVA, L. F. O.; PIO, R.; FIGUEIREDO, M. A.; SILVA, K. N. Establishment of culture medium and quantification of pollen grain germination of quince cultivars. **Revista Brasileira de Fruticultura**, v. 36, n. 2, p. 400-407, 2014.

CHAPTER 4

MEIOTIC STUDIES ON THE GENUS HELICONIA

Species that have not been genetically improved, are still in the wild or are not very domesticated and can show meiotic behaviour with variations between genotypes and many abnormalities in the meiosis and post-meiosis phases, which can lead to limited fertility of male gametes (Damasceno Junior et al., 2010). The observation of non-viable pollen grains may be the result of numerical and structural aberrations resulting from irregularities that can occur during pre-meiosis, meiosis and post-meiosis (Horner & Palmer, 1995).

In view of this, the study of pollen viability together with meiotic behaviour is very important information for planning genetic improvement, as it makes it possible to highlight male reproductive potential and verify the occurrence or absence of observed chromosomal irregularities, which can directly affect the success of crosses between individuals.

Therefore, information on meiotic behaviour, meiotic index and pollen viability is essential for a better understanding of the reproductive aspects of a species, providing information for the development of genetic improvement programmes. Work of this nature has only been carried out on a few species of heliconia.

Meiotic analyses were carried out on the diploid (2n=24 chromosomes) cultivars of *Heliconia psittacorum* 'Tay', 'Andromeda' and 'Lady di', which were found to be fertile, with the formation of regular tetrads, with 99.60%, 99.70% and 100% respectively, indicating normal microsporogenesis. Some microsporocytes divided or remained undivided, giving rise to dyads and monads. However, in the triploid cultivars (2n=36 chromosomes) 'Petra', 'Sassy' and 'Iris' the percentage of tetrads found was 31.30%, 69.60% and 28.50% and the rest were abnormal spores formed, a consequence of their failure to form functional gametes as a result of meiotic irregularities during microsporogenesis and megasporogenesis. As chromosome distribution at meiotic stages is likely to be irregular in triploids, the resulting microspores and megaspores will receive varying numbers of chromosomes (Lee et al., 1994).

In the H. pogonantha species, lagging chromosomes were observed in pollen mother cells in the anaphase I (Figure 15B) and anaphase II (Figure 15A) phases. Cells considered abnormal with a cluster of 5 to 8 microspores (polyads) were also observed (Figure 15C and 15D). Also during the meiosis analyses, empty pollen grains were found that did not show any staining, while full pollens were stained acetic carmine red (Figure 15E). The species showed 35.90% abnormal tetrads made up of polyads, dyads and monads, with the majority being made up of normal tetrads 64.09% (Franco et al., 2012).

In H. psittacorum, pollen mother cells were observed in metaphase I with the 12 bivalents regularly paired (Figure 16A and Figure 16B). However, there were polyads with unbalanced microspores, containing several nuclei or micronuclei (Figure 16C). The pollen grains had a heterogeneous morphology, with some being large, rounded and with numerous ornaments all over the outer wall, and others being small and flatter, with "spicules" only in specific places (Figures 16D and 16E). Pollen grains were also observed with nuclei resulting from the gametogenesis process (Figure 16F) (Franco et al., 2012).

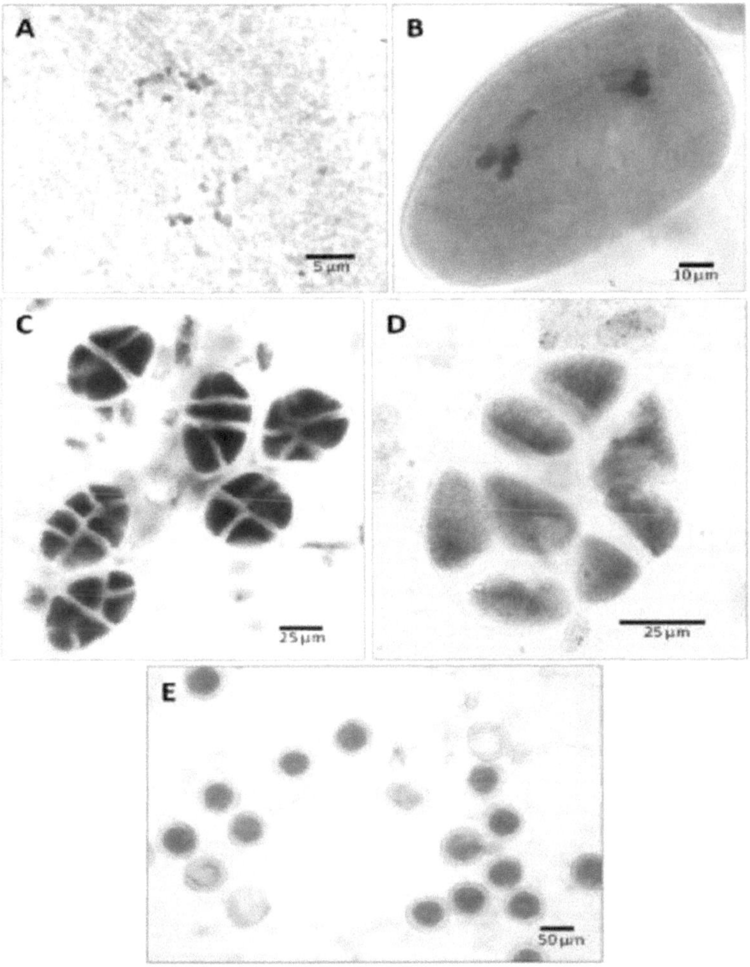

Figura 15. Meiosis and pollen formation in *Heliconia pogonantha:* (A) and (B) anaphase II and anaphase I respectively, both cells showing lagging chromosomes; (C) and (D) sets of polyads with several microspores; (E) viable pollen grains (coloured) and non-viable grains (empty without colour). Coloured with 1.2% acetic carmine. Franco et al., (2012).

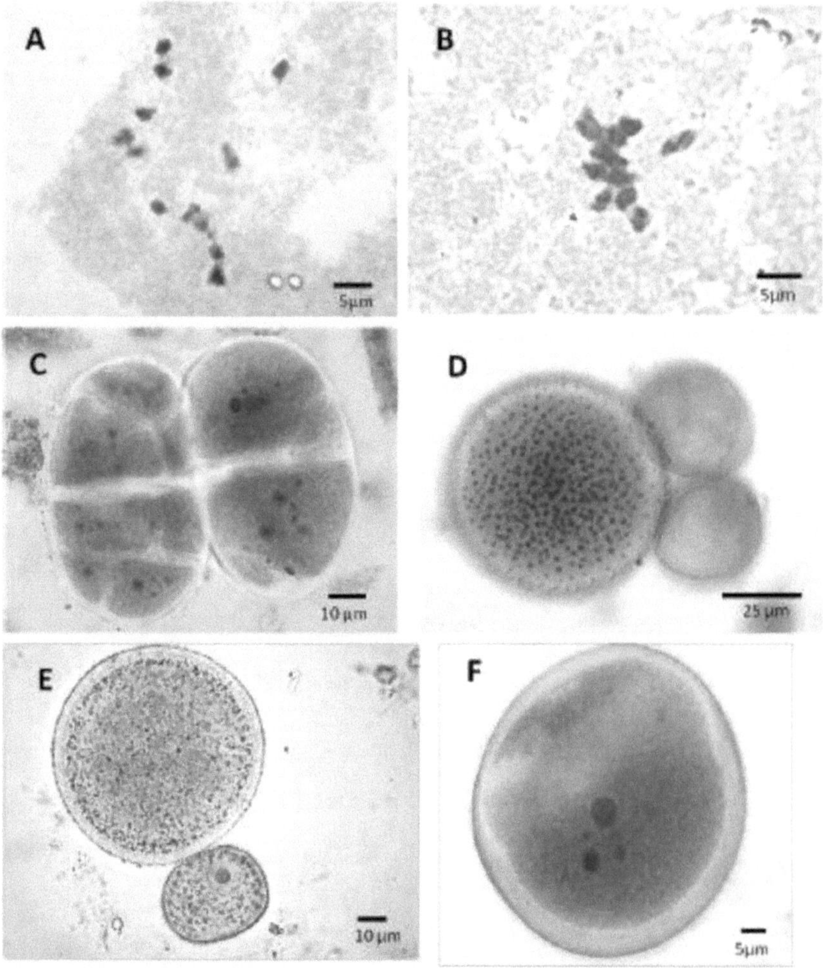

Figura 16. Meiosis and pollen of *Heliconia psittacorum:* (A) Metaphase-anaphase I; (B) Metaphase I; (C) Polyads with several microspores; (D) and (E) Pollen with grains of heterogeneous size; (F) Pollen grain with nuclei resulting from division of gametogenesis. Coloured with 1.2% acetic carmine. Franco et al., (2012).

Bibliographical references

DAMASCENO JUNIOR, P. C.; PEREIRA, T. N. S.; FREITAS-NETO, M.; PEREIRA, M. G. Meiotic behaviour of Carica papaya and Vasconcellea monoica. **Caryologia**, v. 63, n. 1, p. 229-236, 2010.

RANCO, V. R.; ZUCCARI, L. D. A.; PINTO-MAGLIO, C. A. F.; CASTRO, C. E. F. Meiotic behaviour and pollen viability in Heliconia pogonantha and Heliconia psittacorum. In: CONGRESS INTERINSTITUTIONAL SCIENTIFIC INITIATION, 6, 2012, Jaguariúna. **Proceedings...** Jaguariúna: CIIC, 2012. p1-10.

HORNER, H. T.; PALMER, R. G. Mechanisms of genetic male sterility. **Crop Science**, v. 35, n. 6, p. 1527-1535, 1995.

LEE, Y. H.; NG, N. Y.; GOH, C. J. Pollen formation and fruit set in some cultivars of Heliconia psittacorum. **Scientia Horticulturae**, v. 60, p.167-172, 1994.

CHAPTER 5

COLLECTIONS AND GERMPLASM BANKS FOR HELICONIA CONSERVATION

The naturally occurring environments of heliconia species are generally tropical forest regions, along roadsides, riverbanks and in clearings of humid forests, gallery and riparian forests, woodlands and understories, and there are also occurrences in partially open habitats with high solar irradiation (Rundel et al., 1998). Excessive exploitation of these environments, combined with the illegal extraction of species, has reduced the natural reservoir of genes with potential for use present in these resources, leading many species to run the risk of genetic erosion or even extinction.

The destruction of these plants' habitats can isolate them in "forest islands", which are often too small to maintain an acceptable number of individuals of each species. In this way, reducing the population increases the chances of it disappearing, whether due to disease, attacks by herbivores or the destruction of its habitat. In a small population, individuals generally end up interbreeding, and this can lead to genetic problems, which result from crosses between related individuals and also increase the chances of extinction (Costa et al., 2011).

The number of Zingiberales species that may be threatened with extinction is uncertain, simply because little is known about the distribution of each species. The species most at risk of extinction are those that are endemic to regions that have been highly altered by man, such as the tropical rainforest, the habitat that encompasses most of the Zingiberales, and is one of the most threatened environments in the world (Costa et al., 2011). Among the number of known species, 94 are recorded as endemic to a particular region, which shows the genus' great fragility in terms of germplasm conservation (Castro et al., 2007). Many species in the order Zingiberales may probably become extinct before they are even known and described, and measures need to be taken to conserve the species.

One strategy for preserving the Heliconia genus and reducing the risk of genetic erosion is the conservation of these species in germplasm banks, such as ex situ collections, as well as enabling further studies of their floristic and ornamental potential and their use in breeding programmes (Argôlo, 2009). Therefore, germplasm banks are valuable tools for preserving as much of nature's genetic variability as possible, which is vital for scientific plant research and future use by society (Veiga et al., 2012).

The existing collections and germplasm banks of heliconia are worldwide, sparked by the great interest in this genus, which has plants with inflorescences of various colours, shapes and exotic appearances. The Heliconia Society International (HSI), located in Florida (USA), has registered existing collections of Heliconia accessions in its database, with six germplasm banks standing out: Andromeda Garden in western India; Flamingo Gardens in Fort Lauderdale, USA; Harold L. Lyon Arboretum in Honolulu, USA. Lyon Arboretum in Honolulu (Hawaii); Pacific Tropical Botanical Gardens in Kauai (Hawaii); Robert and Catherine Wilson Botanic Garden in Costa Rica, and Jurong Bird Park in Singapore, as well as The Marie Selby Botanical Gardens in Florida (Kress, 1987; Bannochie, 1987; Lorence, 1999), which can be accessed through the National Tropical Botanical Garden, considered the official conservation centre of the HSI, dedicated to research and conservation of tropical plants with more than 200 heliconia accessions.

Existing germplasm banks in Brazil are generally found in educational and research institutions, by collectors and in gardens (Costa, 2013). The establishment of heliconia conservation collections began in 1987 at the Roberto Burle Marx site in Rio de Janeiro. The main banks are kept at the Agronomic Institute of Campinas-IAC, with 180 accessions of 60 species, at the Botanical Institute of the State of São Paulo, at EMBRAPA centres, at Federal Universities and in some Botanical Gardens (Berry, 1989; Castro, 2009).

Some public universities also have heliconia collections. The Federal Rural University of Pernambuco (UFRPE) in Pernambuco has a collection located at Fazenda-Bem-Te-Vi in the municipality of Camaragibe at Aldeia - PE, which began in 2003 and includes genotypes acquired from producers in various states: Pernambuco, Alagoas, Bahia, Rio de Janeiro, São Paulo and Pará, containing around 50 genotypes (Costa, 2013). The active germplasm bank (BAG) of the State University of Mato

Grosso (UNEMAT) located in the experimental field of the Tangará da Serra-MT Campus, set up in March 2014, has 30 heliconia accessions collected in different municipalities in the state of Mato Grosso. The H. psittacorum and H. densiflora species stand out for having the greatest diversity of accessions and variations in the colour of the bracts.

In addition, the Ubatuba Research and Development Unit (UPD), linked to the Paraíba Valley Regional Centre/APTA since 1931, has been involved in research and development of tropical plants. The "Introduction and Selection of Ornamental Zingiberals" project is being implemented at the unit, including the heliconia germplasm bank project, which has 180 accessions of 60 species and several varieties (Vega et al., 2011). In addition to conserving these species, the Unit is implementing a project to distribute collections of heliconia species, with the aim of disseminating ornamental plants with current or potential economic value and subsidising research, combined with the growing demand for tropical flowers with the region's tourist potential and preservationist requirements.

Bibliographical references

ARGÔLO, L. M. H. **Evaluation of** Heliconia **spp. genotypes under full sun and cabruca cultivation.** 76f. Dissertation (Master's in Plant Production) - State University of Santa Cruz, Ilhéus, 2009.

BANNOCHIE, I. Attribute to the Zingiberaceae. **Heliconia Society International Bulletin**, v.2, n. 3/4, p.7-9, 1987.

BERRY, F. Heliconia at Sítio Roberto Burle Marx. **Bulletin** Heliconia **Society International**, v.4, n.2, p.7, 1989.

CASTRO, C. E. F.; GONÇALVES, C.; MAY, A. Updating the nomenclature of species of the genus Heliconia (Heliconiaceae). **Revista Brasileira de Horticultura Ornamental**, v.13, p. 38-62. 2007.

CASTRO, C. E. F. History of Brazilian floriculture: organisation and evolution. **Magistra**, v. 21, n. 4, p. 233-252, 2009.

COSTA, F. R. C.; ESPINELLI, F. P.; FIGUEIREDO, F. O. G. **Guia de zingiberales dos Sítos PPbio na Amazônia Ocidental brasileira**. Manaus: Áttema Design Editorial, 2011. 284p.

COSTA, M. G. S. **Cytogenetic characterisation of genotypes of** Heliconia **L. (Heliconiaceae).** 54f. Dissertation (Master's Degree in Agronomy - Plant Genetic Improvement), Federal Rural University of Pernambuco, Recife, 2013.

KRESS, W. J. Heliconia research at the Marie Selby Botanical Gardens. **Bulletin Heliconia Society International**, v.3, n.1, p. 1, 1987.

LORENCE, D. H. Zingiberales collection at the National Tropical Botanical Garden: 1998 report. **Bulletin** Heliconia **Society International**, v. 9, n. 4, p.8-9, 1999.

RUNDEL, P. W.; SHARIFI, M. R.; GIBSON, A. C.; ESLER, K. J. Structural and physiological adaptation to light environment in neotropical Heliconia (Heliconiaceae). **Journal of Tropical Ecology**, v.14, p. 789-801, 1998.

VEIGA, R. F. A.; BARBOSA, W.; TOMBOLATO, A. F. C.; MONTENEGRO, J. F. V. Germplasm banks: importance and organisation. In: COSTA, A. M.; SPEHAR, C. R.; SERENO, J. R. B. **Conservation of genetic resources in Brazil**. Brasilia: EMBRAPA, 2012. p. 105.

VEGA, S. M. R.; CASTRO, C. E. F.; MARCHIORI, A. C. C.; GONÇALVES, C. A research, innovation and technology transfer in the UPD of Ubatuba. **Pesquisa & Tecnologia**, v. 8, n. 2, 2011.

CHAPTER 6

THE MARKET FOR TROPICAL ORNAMENTAL PLANTS AND THE USE OF HELICONIA IN PLANT BREEDING

Tropical flowers have characteristics that set them apart from the rest, such as their exotic appearance, contrasting colour and post-harvest longevity, which favour their acceptance and interest by consumers, making them a promise of satisfactory profitability for many producers (Neto, 2010).

In both the national and international ornamental plant markets, species of tropical origin account for a portion of the interest in these plants. Among the most important tropical flowers grown in Brazil are the genus Heliconia L., the genera Alpinia and Etlingera (Emperor's Bastion), Costus, both from the Zingiberaceae family, and the genus Anthurium (Araceae) (Lamas, 2004). Figure 17 shows the species that are currently the most commercialised internationally. Followed by the species most produced in Brazil (Figures 18 and 19).

Although several species in the Heliconia genus have favourable prospects for use as cut flowers, few have been cultivated for this purpose (Castro et al., 2007). They show great variability, capable of being exploited by genetic improvement programmes, but research into the pre-breeding and genetic improvement of these plants is scarce in the country, and they are also restricted to a few institutions and companies that carry out studies in this area (Rocha, 2009). By characterising these species and their potential use, their cultivation can be increased by rural producers, favouring information that expands their use, increasing the supply on the market.

Figure 17. Most internationally traded species: (A) *H. stricta* cv. Dwarf Jamaican (B) **H.** *stricta* cv. Sharonii, (C) **H.** *bihai* cv. Lobster Claw (D) **H.** *bihai* Nappi Yellow (E) **H.** *bihai* Dwarf, (F) **H.** *orthotricha* cv. Edge of Nite, (G) **H.** *orthotricha* cv. She, (H) H. *psittacorum* cv. Andromeda, (I) **H.** *psittacorum* cv. Sassy, (J) **H.** *psittacorum* St. Vincent Red, (K) **H.** *angusta* cv. Holiday, (L) **H.** *angusta* Orange Chistmans, (M) **H.** *chartacea* cv. Sexy Pink, (N) *H. rostrata,* (O) *H. xanthovillosa,* (P) *H velerigera,* (Q) *H. wagnoriana,* (R) *H. velloziana,* (S) *H. psittacorum* x *H. spathocircinata* cv. Golden Torch. Heliconia Society of Puerto Rico, 2017.

According to Criley (2000), heliconias collected in forests are generally introduced with little or no information about their flowering periods and are therefore often unsuitable for cultivation and commercialisation. Therefore, it is extremely important to collect agronomic information during the evaluation of introduced material, such as characteristics such as: beginning of tillering, beginning of flowering, production of floral stems, length of floral stems, length of inflorescence, fresh mass of floral stems, shape of inflorescence, cycle, period for issuing inflorescence, among others (Rocha, 2009).

Figure 18. Most produced species in Brazil: (A) and (B) *H. bihai;* (C) and (D) *H. wagneriana;* (E), (F) and (G) **H.** *stricta* and (H) H. *psittacorum* x **H.** *spathocircinata* cv. Golden Torch (Castro et al., 2007). National Tropical Botanical Garden, 2017.

Figure 19. *Heliconia psittacorum* L.f. one of the most cultivated species in Brazil, with small, light bracts and different colours and shapes of inflorescences (A) red, (B) orange-red, (C) dark pink and (D) yellow.

Some aspects are fundamental to the development of a breeding programme, among them: research into the diversity and variability of the species, carrying out the morphological, cytogenetic, biochemical, molecular and agronomic characterisation stages; determining the breeding methods to be used; financial planning; surveying the demands of the consumer market and the commercial importance of the end product (Pereira, 2010).

Genetic improvement programmes aimed at the production of *Heliconia* should take into account a number of characteristics, including: beauty, productivity, growth, vigour, post-harvest durability, size and shape during packaging (Criley and Broschat, 1992). In addition to obtaining information on flowering time, this must be obtained before introducing new material, since the time and period of flowering directly influence the application of the methods to be used in breeding (Pereira, 2010).

Knowledge of these characteristics is important for the selection and introduction of new materials. In Colombia, a heliconia research programme was created due to the great natural diversity. The programme

was aimed at varietal selection of genetically superior plants adapted to different environments. It also aimed to reduce the size and weight of the stems to facilitate transport; to develop types with characteristics desirable to the consumer; to increase longevity and reduce the production of ethylene, which damages the stems during transport and storage (Atehortua, 1997).

Another important characteristic to be assessed and identified is disease resistance in heliconia species, which could be used in breeding programmes aimed at resistance, creating hybrids that can be established in regions and conditions that favour disease attack, but which are resistant.

Bipolaris blotch, anthracnose, cercospora blotch, rust, phytonematosis, bacteriosis, rhizome rot are some of the diseases that occur in heliconia (Assis et al., 2002; Gasparotto et al., 2005; Kimati et al., 2005; Lins & Coelho, 2004). Silva et al. (2015) verified the occurrence of phytopathogens that cause fungal diseases: Colletotrichum gloeosporioides, Fusarium sp., Curvularia lunata, and Bipolaris sp. In addition to these, Cercospora sp., Cladosporium herbarum, Nigrospora sp. and Helminthosporium sp. were also identified, highlighting the diversity of phytopathogens that cause fungal diseases in heliconia (Nascimento et al., 2015).

The symptoms of these diseases include dark spots and dryness on the leaves and inflorescences, which cause damage to the crop, such as a reduction in the photosynthetic area, affecting the yield of marketable stems. In this way, surveying the occurrence of diseases that attack heliconia can generate information for farmers and producers on the management of these plants, helping them to choose appropriate control methods, reducing the damage caused to the plant and production (Nascimento et al., 2015).

The genetic improvement programmes for heliconia must meet the demands of the consumer market, since its cultivation has shown significant growth in recent years, establishing a promising market for the national producer of ornamental plants, with some species being cultivated intensively to meet the market for cut flowers or plants for gardens (Castro et al., 2007), 2007), so the creation of germplasm collections and banks is essential for conserving the variability of heliconia species in the long term, so that in the future they can be used in research and genetic improvement, in order to obtain new cultivars that cater for the ornamental plant consumer market.

Bibliographical references

ASSIS, S. M. P.; MARINHO, R. R. L.; GONDIM JUNIOR, M. G. C.; MENEZES, M.; ROSA, R. C. T. **Diseases and pests of heliconia**. Recife: UFRPE, 2002. 102p.

ATEHORTUA, L. Heliconias: A new challenge for the Colombian floricultural industry. **Biotechnology and Development Monitor**, v.31, p.2021, 1997.

CASTRO, C. E. F.; GONÇALVES, C.; MAY, A. Updating the nomenclature of species of the genus Heliconia (Heliconiaceae). **Revista Brasileira de Horticultura Ornamental**, v.13, p. 38-62. 2007.

CRILEY, R. A.; BROSCHAT, T. K. Heliconia: botany and horticulturae of new floral crops. **Horticulturae Review**, v. 14, p.1-55, 1992.

CRILEY, R. A. Seasonal flowering patterns for Heliconia shown by grower records.

Acta Horticulturae, v.541, p.159-165, 2000.

GASPAROTTO, L.; PEREIRA, J. C. R.; URBEN, A. F.; HANADA, R. E.; PEREIRA, M. C. N. Heliconia psittacorum: Host of Mycosphaerella fijiensis, Causal Agent of Banana Black Sigatoka-. **Fitopatologia Brasileira**, v.30, p.423-425, 2005.

KIMATI, H.; AMORIM, L.; REZENDE, J. A. M.; CAMARGO, L. E. A. **Manual de fitopatologia: doenças de plantas cultivadas**. 4.ed. São Paulo: Editora Ceres, 2005.

LAMAS, A. M. **Tropical Floriculture: cultivation techniques.** Recife: SEBRAE-PE, 2004. 65p.

LINS, S. R. O.; COELHO, R. S. B. Occurrence of diseases in tropical ornamental plants in the state of Pernambuco. **Fitopatologia Brasileira**, v.29, p.332-335, 2004.

NASCIMENTO, T. O.; CARDOSO, V. L. J. L.; SILVA, P. C.; SILVA, C. G.; DALBOSCO, E. Z.; HIEGA, K. M. R.; ARAÚJO, D. V. Fungal diseases in Heliconia in the municipality of Tangará da Serra - MT. **Revista MT Horticultura**, v. 1, n. 2, p. 08-11, 2015.

NETO, A. S. M. **Characterisation and economic viability of the tropical flower production system in Rio de Janeiro**. 86f. Dissertation (Master's in Plant Production), State University of Northern Rio de Janeiro Darcy Ribeiro. Campos dos Goytacazes, 2010.

PEREIRA, T. N. S. **Germplasm: conservation, management and use in plant breeding**. Viçosa: Arca, 2010. 250p.

ROCHA, F. H. A. **Agronomic characterisation, variability, correlations and repeatability in** Heliconia psittacorum **cultivars and interspecific hybrids**. 58f.

Dissertation (Master's Degree in Agronomy - Plant Genetic Improvement) - Federal Rural University of Pernambuco, 2009.

SILVA, G.; HIEGA, K. M. R; DALBOSCO, E. Z.; SILVA, C. A.; ARAÚJO, D. V. Plant health in tropical plants in the state of Mato Grosso. **Enciclopédia biosfera, Centro Científico Conhecer**, v.11 n.22, p.1293, 2015.

CHAPTER 7

BIOTECHNOLOGICAL STUDIES: IN VITRO CULTIVATION AND POLYPLOIDISATION OF HELICONIAS

The cultivation of heliconia seedlings has faced a serious problem that directly influences the production potential of these plants. Propagation by seed is time-consuming, delaying the flowering period, which begins three or four years after the start of vegetative growth. Propagation through rhizomes, which is the most widely used, becomes a means by which diseases are spread on a large scale in plantations. Biotechnological studies of in vitro culture with heliconia are therefore being carried out in order to solve this problem.

In vitro micropropagation techniques for ornamental plants provide accelerated production development, guarantee a reduction in the physical space used and anticipate the plant's development process (Bortolin, 2006). In addition to these advantages, this technique can be combined with mutagenesis and polyploid induction methods aimed at genetic improvement, allowing for an increase in the variability of plants with ornamental potential, which is gaining more and more ground in all regions of the country (Sleper & Poehlman, 2006).

The use of in vitro propagation for heliconias in conjunction with other techniques can be used for the conservation and evaluation of germplasm and for use in breeding programmes. By applying doses of growth regulators and incubation periods to the rooting and initial sprouting of rhizomes of Heliconia psittacorum l.f. x Heliconia spathocircinata aristeguieta cv. alan carle. It was found that vegetative propagation for both species is more efficient with the use of the growth regulator AIB (indolebutyric acid) at a concentration of 10g/l, and its use combined with an incubation period of 48 hours makes this application even more efficient (Melo, 2011).

When colchicine was applied to Heliconia bihai (L) L. to induce polyploidy in vitro, two individuals were polyploidised at concentrations of 0.05% and 0.1% colchicine, both for 48 hours at the tetraploid level (Figures 20 and 21). The data was confirmed by analysing the stomata (Figure 22), which showed changes in size and number, and by analysing the number of chromosomes. The author found 2n = 48 chromosomes for the 0.05% treatment (16%) and for the 0.1% treatment for 48 hours (16%), both considered possible tetraploids (4x) (Figure 23). The results indicate that the induction of polyploids is viable for H. bihai, with the concentrations of 0.05% and 0.1% for 48 hours being ideal for achieving this result (Filho, 2011) (Figure 20).

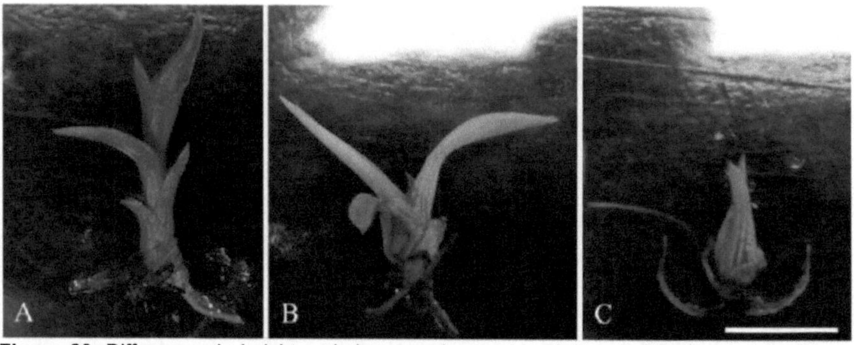

Figure 20. Differences in height and changes observed in **Heliconia bihai** material subjected to different concentrations of colchicine **in vitro**, at 30 days. A: Control material. B: Material subjected to 0.05% colchicine for 48 hours, showing long, thin leaves and swelling at the base. C: Explant treated with 0.1% colchicine for 24 hours, showing slow development. Scale = 1 cm. Filho, 2011.

Figura 21. Mitotic metaphase chromosomes *of Heliconia bihai* plants. *A:* Diploid control (2n = 24 chromosomes). B: Probable tetraploid (2n = ca. 48 chromosomes) subjected to 0.05% colchicine for 48 h; C: Probable tetraploid (2n = ca. 48 chromosomes) subjected to 0.1% colchicine for 48 h. Scale = 10 µm. Filho, 2011.

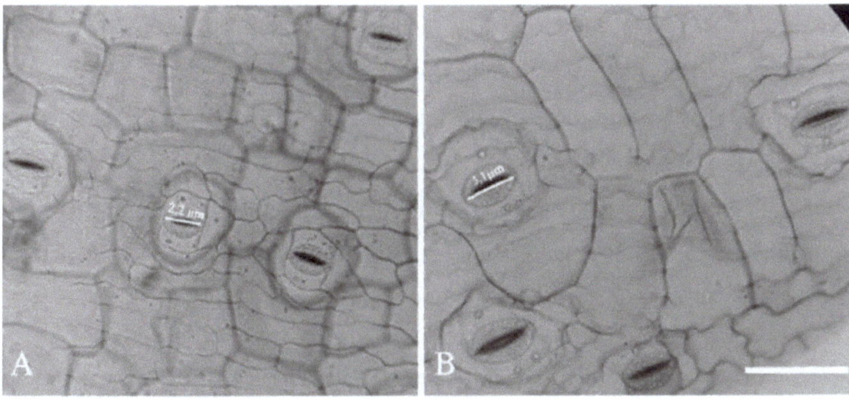

Figura 22. Stomata size of *Heliconia bihai* plants subjected to different concentrations of colchicine *in vitro.* A: Diploid control material. B: Polyploid material subjected to 0.05% colchicine for 48 hours. Scale = 5 µm. Filho, 2011.

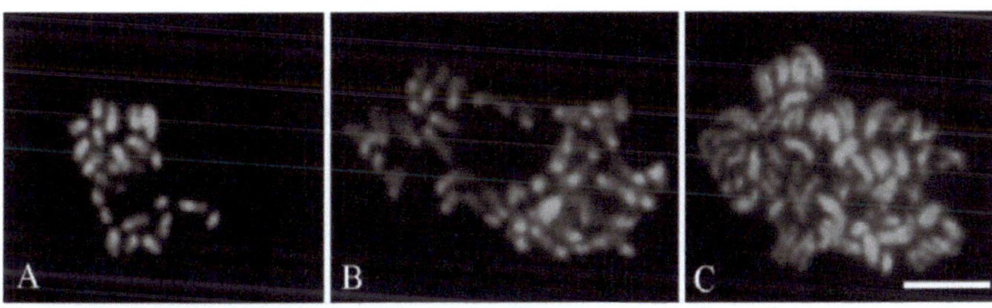

Figure 23. *Heliconia bihai* plants subjected to different concentrations of colchicine *in vitro*, at 120 days. A: Control material. B: Polyploid (2n = 48 chromosomes) subjected to 0.05% colchicine for 48 h; C: Polyploid (2n = 48 chromosomes) at 0.1% 48 h. Scale = 2 cm. Filho, 2011.

The species *H. ortrotricha* cv. Candy Cane provided significant multiplication during *in vitro* cultivation, but *H. bihai* cv. Peach Pink and *H. ortrotricha* L. Anderss. cv. Total Eclipse showed a very low rate of sprouting. The Murashige and Skoog (1962) medium, with a normal salt concentration, plus 2.5 mg/L BAP and 0.10 ANA was the most appropriate medium for growing *H. ortrotricha* cv. Candy Cane in vitro, considering the number of shoots and the general characteristics of the plants (Nakano, 2008). It should be noted that tests to determine the conditions required for *in vitro* multiplication of heliconia are necessary, as the cultivation requirements vary from species to species (Figure 23).

Torres et al. (2005) evaluated the development of disease-free *Heliconia rostrata* seedlings through embryo culture and observed that sucrose added to the culture medium was effective for embryo development, since in its absence the embryos did not develop and died. Concentrations of 1%, 2% and 3% (w/v) sucrose allowed the embryos to develop, while concentrations of 6%, 9% and 12% (w/v) sucrose inhibited their growth. The addition of kinetin, isopentenyl adenine and zeatin did not favour the growth and development of the embryos. The authors also emphasise the importance of this study due to the ineffectiveness of propagation methods

for H. rostrata, favouring production with reduced costs in obtaining genotypes of high phytosanitary quality, in the introduction of accessions into in vitro germplasm banks and in the genetic improvement of this species.

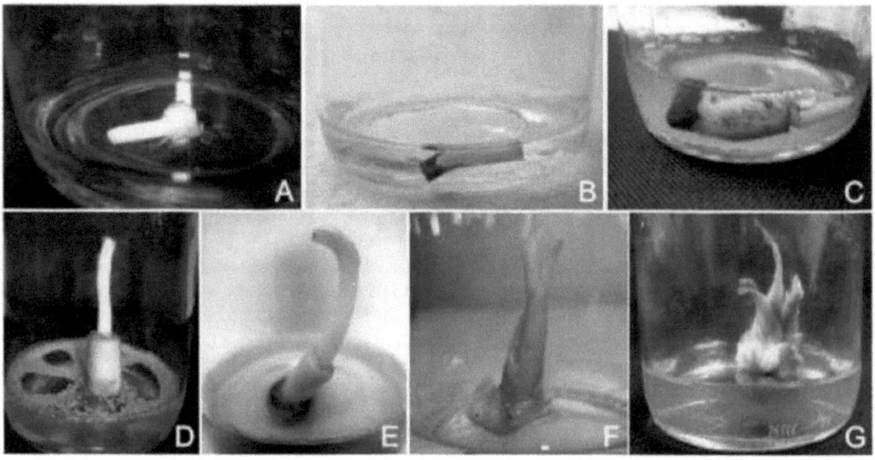

Figure 24. *In vitro* introduction of heliconia stem apices: A) Heliconia stem apex introduced in *vitro* in liquid MS medium; B) Appearance of explant 7 days after introduction *in vitro;* C-D) Bacterial contamination of explant in liquid (C) and semi-solid (D) culture medium; E) Regeneration of heliconia stem apex established *in vitro.* F) Characteristic of the explant after 45 days *in vitro* culture. G) Explant showing sprouts after 60 days of *in vitro* establishment of the stem apex. Nakano, 2011.

Bibliographical references

BORTOLIN, B. Flowers: high-tech production and diversification. **Inovação Uniemp,** Campinas, v.2, n.1, 2006.

FILHO, G. J. F. C. **Induction of Polyploidy** in vitro **with colchicine application in**

Heliconia bihai **(L) L.** 63f. Dissertation (Master's Degree in Biological Sciences) - Federal University of Pernambuco, Recife, 2011.

MELO, A. A. **Effect of applying doses of growth regulator and incubation periods on the rooting and initial sprouting of rhizomes of** *Heliconia psittacorum* **l.f. x** *Heliconia spathocircinata* **aristeguieta cv. alan carle**. 46f. Dissertation (Master's Degree in Agronomy and Veterinary Medicine) - University of Brasília, Brasília, 2011.

MURASHIGE, T.; SKOOG, F. A revised medium for rapid growth and bioassays with tissue cultures. **Physiologia Plantarum**, v.15, p. 473-497, 1962.

NAKANO, V. A. **Micropropagation of heliconia species, morphological characterisation and molecular identification of contaminating bacteria.** 81f. Dissertation (Master of Science), University of São Paulo, Piracicaba, 2008.

SLEPER, D. A.; POEHLMAN, J. M. **Breeding field crops**. Ames, Iowa, U.S.A.: Blackwell Publishing, 2006, 424 p.

TORRES, A. C.; DUVAL, F. G.; RIBEIRO, D. G.; SANTOS, M. D. M. **Production of Disease-Free** *Heliconia rostrata* **Seedlings via Embryo Culture.** Brasília: Embrapa Hortaliças, 2005. 12 p.

MIX
Papier aus verantwortungsvollen Quellen
Paper from responsible sources
FSC® C105338

Printed by Books on Demand GmbH, Norderstedt / Germany